Water 113

# 特殊的爸爸

# My Special Dad

*Gunter Pauli*

[比] 冈特 · 鲍利 著

[哥伦] 凯瑟琳娜 · 巴赫 绘

郭光普 译

上海遠東出版社

特别感谢以下热心人士对童书工作的支持：

匡志强　宋小华　解　东　厉　云　李　婧　庞英元
李　阳　梁婧婧　刘　丹　冯家宝　熊彩虹　罗淑怡
旷　婉　王靖雯　廖清州　王怡然　王　征　邵　杰
陈强林　陈　果　罗　佳　闫　艳　谢　露　张修博
陈梦竹　刘　灿　李　丹　郭　雯　戴　虹

# 目录



# Contents

一只海马正在进行一趟稀罕的北极之旅。这是一个美好的夏日，但是对海马来说，他觉得自己冷得快要结冰了。这时，他遇见一只正在筑巢的鸟。

“你好，小鸟夫人，请问你能告诉我是谁教会你筑巢的吗？”海马问。

A seahorse is making a rare trip to the Arctic. It is a beautiful summer’s day, but for the seahorse it is freezing cold. He meets a bird making a nest.

“Hello Mrs Bird. Could you please tell me who taught you to build a nest?” Seahorse asks.

一只海马正在进行一趟北极之旅。

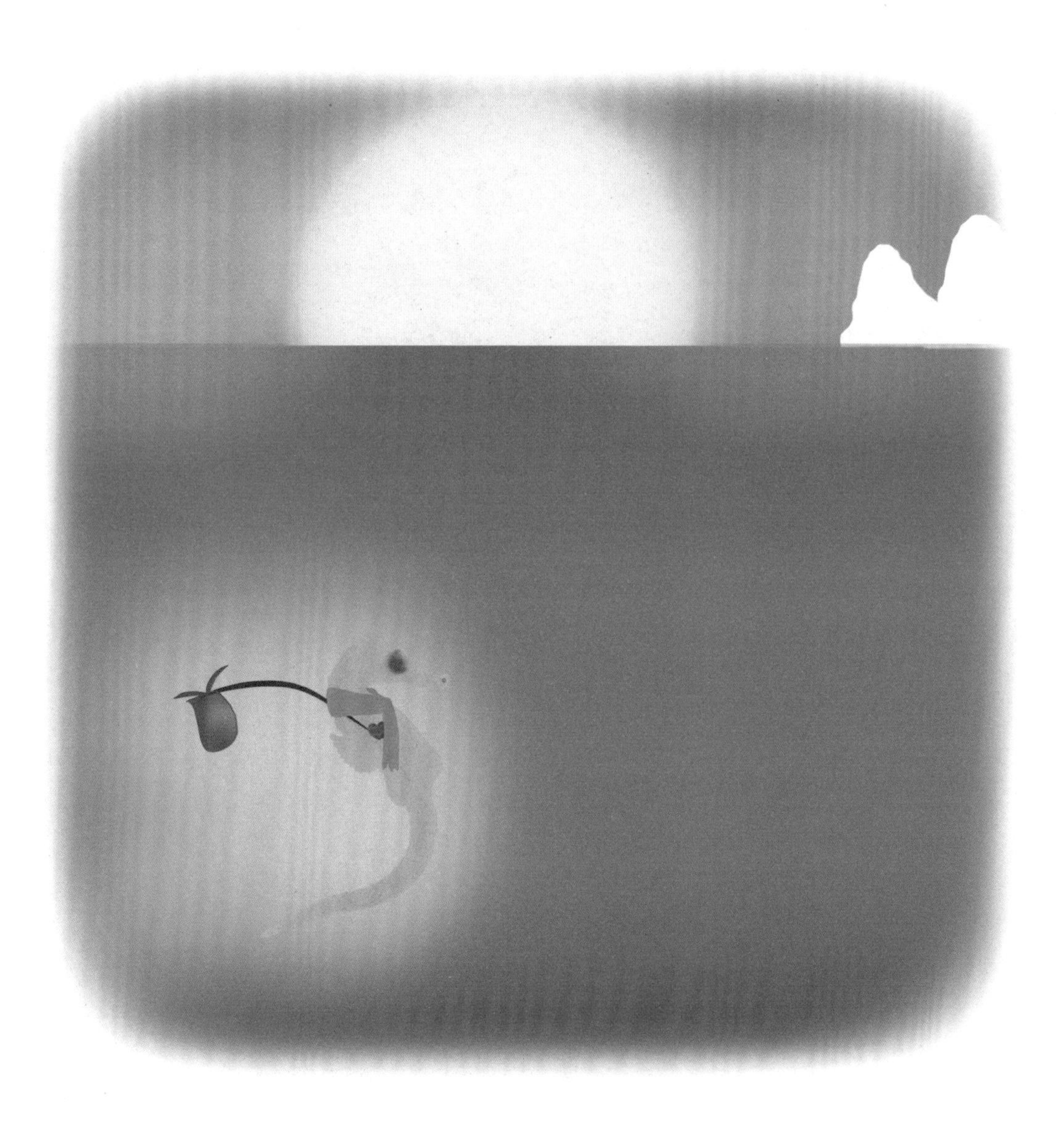

A seahorse is making a trip to the Arctic.

我是这个新家庭未来的父亲。

I am the future father of a new family.

“哦，我不是这房子的女主人！事实上，我是这个新家庭未来的父亲。”这只瓣蹼鹬回答道。

“我很抱歉！我看到你的伴侣长得更大，并且颜色更鲜艳，所以我还以为你是妈妈，而那位是爸爸。”

“Oh, I am not the lady of the house! I am in fact the future father of a new family,” the phalarope responds.

“I do apologise. I saw that your partner is bigger and more brightly coloured, so I thought you are the mom, and that one the dad.”

“没关系。”瓣蹼鹬回答，“你知道吗，在北极的鸟类世界中，女士们为了我们男士而互相争斗。在寻找伴侣这方面，我们似乎打破了陈规。”

“真的吗？而你负责筑巢？”

“Not to worry,” Phalarope replies. “You know that here in the Arctic world of birds the ladies fight each other for a male like me. When it comes to finding partners, we seem to break the stereotype.”

“Really? And you build the nest?”

……女士们为了我们男士而互相争斗……

... the ladies fight each other for a male ...

……还负责喂养他们……

... also in charge of rearing the nestlings ...

“是的，我筑巢时很快乐。在小鸟宝宝孵化后，我还负责喂养他们。”

“你真是太棒了。我相信很多妈妈和孩子都希望能拥有你这样一位父亲。”

“Yes, and I do it with pleasure. I am also in charge of rearing the nestlings, from the minute they hatch.”

“That is very kind of you. I am sure many moms and kids would love to have a dad like you around.”

“嗯，我要确保我们的宝宝得到很好的照顾，直到他们离巢。当我们向温暖的南方迁徙的时候，他们的妈妈会早些离开，而我留下来照顾孩子，直到他们羽翼丰满，能够飞行。”

“既然你忙于照顾孩子，那是不是妈妈负责寻找食物呢？”

“Well, I want to make sure that our chicks are well cared for until they leave the nest. When it is time to migrate to the warm south and their mom leaves us early, I stay behind and rear the nestlings until they are fledglings, ready to fly.”

“If you are so busy with the chicks, is their mom taking care of finding food?”

……照顾他们一直到他们能够飞行……

... rear the nestlings until they can fly ...

我绕着小圈游泳。

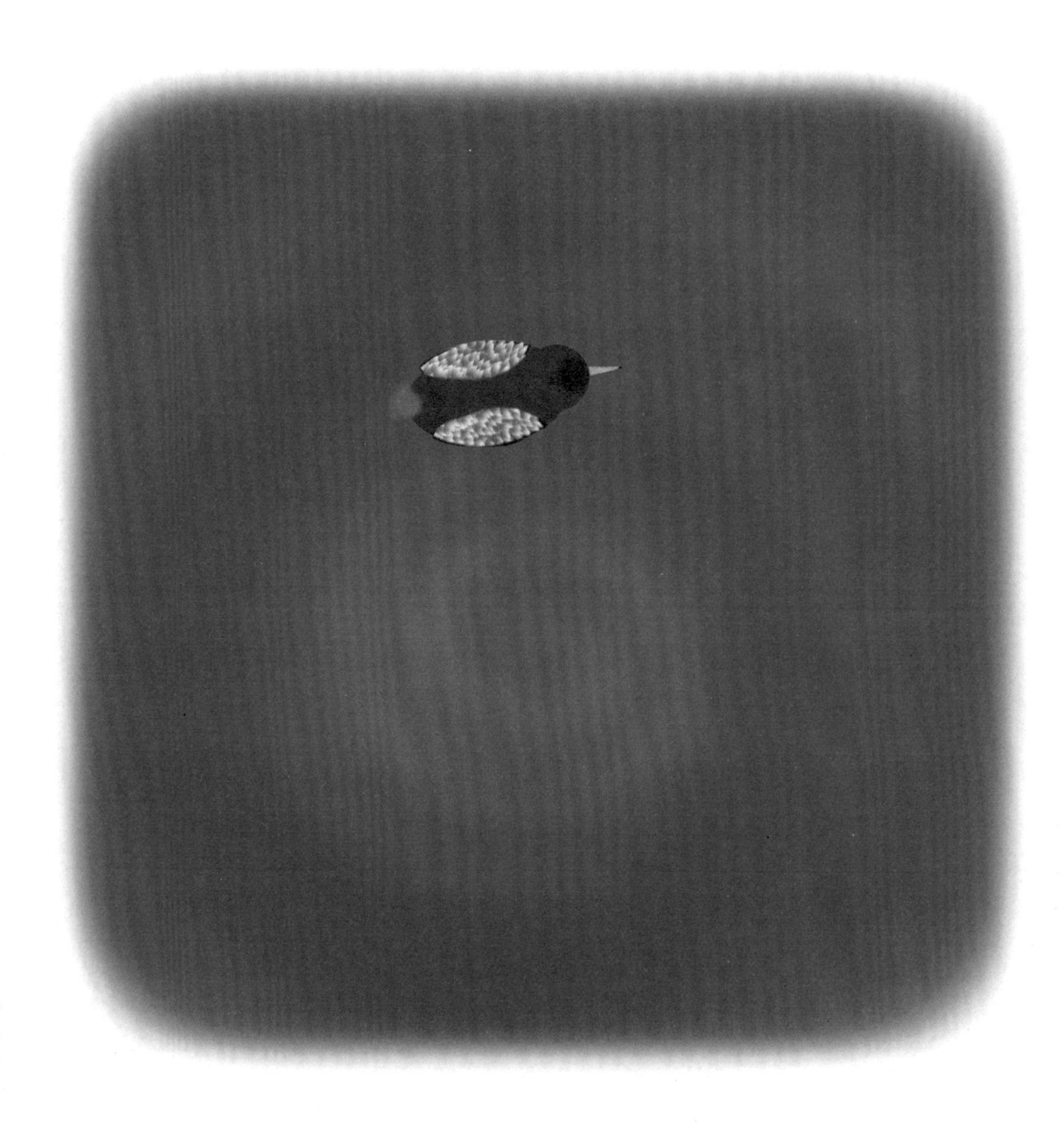

I swim in small circles.

“这个嘛，妈妈总是乐意提供力所能及的帮助。但是我有特殊的办法能迅速找到食物。”

“特殊的办法？”海马问道。

“也没有什么奇特的啦，我只是绕着小圈游泳。”

“Well, Mom is always happy to help where she can. But I have a special technique to find food fast.”

“Some special technology?” Seahorse asks.

“No, nothing fancy. I just swim in small circles.”

“绕着圈游泳是怎么让你更快找到食物的呢？”

“这样可以搅出小漩涡，把食物从水底卷到水面。”

“好聪明！当我们养育孩子时，我们需要所有能得到的帮助。”

“And how does going around in circles help you find food faster?”

“My little whirlpool draws food up, from the bottom to the surface.”

“Now that is smart! We need all the help we can get when we raise our young.”

……小漩涡把食物卷上来……

… whirlpool draws food up …

……富含食物的水流直接冲到嘴里。

... water that is full of food flow directly up my bill.

“我还可以让富含食物的水流直接冲到我的嘴里。”

“听起来像个奇迹！即使我不明白你究竟是怎么做到的。”

“我听说你也能创造奇迹呀。”瓣蹼鹬说，“你——作为父亲——可以生出小海马，这是真的吗？”

“I am also able to make water that is full of food flow directly up my bill.”

“That sounds like a miracle! Even if I do not understand how you do it.”

“But I hear that you also perform miracles,” Phalarope says. “Is it really true that you – the dad – give birth to your little ones?”

“是的。真遗憾世界上没有更多像我们这样特殊的爸爸。我多想看到更多特殊的爸爸呀！”

“这世上没有那么多富有爱心和关怀体贴的父亲吗？这也是一种刻板印象，不是吗？我觉得这样的父亲比人们认为的要多得多。”瓣蹼鹬笑着说。

……这仅仅是开始！……

“Yes, it is. What a pity there are not more exceptional dads like us in the world. I would love it to see more special dads!”

“Not enough loving and caring fathers in the world? Well, that is a stereotype as well, isn’t it? I think that there are many more than people think,” Phalarope says with a smile.

... AND IT HAS ONLY JUST BEGUN! ...

……这仅仅是开始！……

… AND IT HAS ONLY JUST BEGUN! …

# Did You Know?

Birds build nests with remarkable strength and durability out of local materials that are not particularly strong. The interwoven and entangled composite structures are made from grass, leaves, twigs, bones, wool, mud, and spider silk, and even plastics.

鸟类可以就地取材建造出十分坚固耐用的鸟巢。这些交织缠结的复合结构由草、叶、枝、骨、羊毛、泥和蜘蛛丝，甚至塑料制成。

The large-scale platform nests built by eagles can weigh as much as two tons. These structures are built on high ledges, in trees and even on telephone poles. The nests are reused and the birds will repair the nest for the next clutch of eggs.

老鹰建造的大型平台巢穴可重达两吨。它们将这些建筑物建在悬崖、树木甚至电线杆上。这些巢穴能重复利用，老鹰们会修补它们来迎接下一窝鸟蛋。

Birds like the orioles make hanging nests, far out on a limb, away from predators. Orioles weave plant material with their beaks, and even tie knots to hold the nest together.

一些鸟类，比如黄鹂，会在树枝最末端建造悬垂的巢来躲避捕食者。黄鹂用它们的喙编织植物材料，有时甚至能打结，来保持巢的牢固。

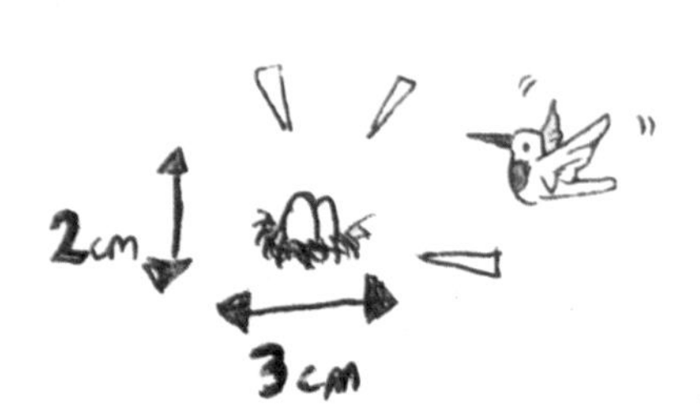


The smallest bird nests are those made by hummingbirds, creating tiny cups of only 2 cm high and 2-3 cm across. The dusky scrubfowl engineers a nest with a diameter of 11 metres and up to 5 metres tall.

最小的鸟巢是由蜂鸟建造的，只有 2 厘米高，2—3 厘米宽，呈小杯子形。暗色冢雉则可以建造出直径达 11 米、高达 5 米的巨大鸟巢。

Hamerkops assemble thousands of twigs into a nest that is insulated and made waterproof with mud. The horned coot makes its nest on top of an island of pebbles that it piles up in shallow lakes. The edible nest swiftlet builds its home entirely from its own saliva.

锤头鹳用成千上万的树枝组装成一个有隔热功能的巢，还在外面糊上泥巴来防水。角骨顶把巢建在其用鹅卵石在浅湖中堆出的小岛上。可以食用的燕窝则是由金丝燕自己的口水建成的。

Seahorses eat 30 to 50 times per day, mainly small crustaceans and shrimp. A baby seahorse (called fry) eats a staggering 3,000 pieces of food per day. A male seahorse stays within 1 square metre of habitat, while a female's range is about one hundred times that of the male.

海马每天要进食 30—50 次，主要摄食小型甲壳纲动物和虾类。小海马（被称为幼苗）每天要吃 3 000 块食物，数量惊人。雄性海马的栖生地面积只有 1 平方米，而雌性海马的栖生地面积则是雄性的 100 倍左右。

Seahorses don't have scales but rather have a skin, and their eyes move like those of a chameleon. Seahorses are slow swimmers, with some of them moving only 1.5 m per hour.

海马没有鳞片，但有一层皮肤，而且它们的眼睛可以像变色龙的眼睛一样运动。海马的游泳速度十分缓慢，有些海马每小时只能移动1.5米。

The seahorse has a reversed pregnancy. The female transfers her eggs to the male, which he self-fertilises in his pouch. The larger seahorse can have up to 1,500 eggs. When giving birth, contractions can last up to 12 hours.

海马在怀孕这件事上会出现角色互换的现象。雌性海马将其卵子转移给雄性，而雄性在其育儿袋里完成自体授精。较大的海马一次可以产1 500个卵。当它们分娩时，育儿袋收缩时间可以长达12小时。

Do you think that there are always exceptions to the rule?

你认为所有的规则都有例外吗?

Is an exception a sign of flexibility, or is it a proof that the rule is too rigid?

例外是灵活性的标志，还是规则太死板的证明?

Why would birds not all make the same type of nest? Why are there so many different ways of achieving the same end: a safe place for eggs to hatch?

为什么鸟类不会建造一样的巢呢?为什么有如此多的方法来达到同一个目的：有一个安全的地方来孵蛋?

Does it make sense for a father to give birth to babies?

父亲生育小宝宝合理吗?

## Do It Yourself!

## 自己动手！

Have a look at how you may perhaps stereotype people. Do you have a fixed characterisation of all sportsmen being rough, or rock musicians leading undisciplined lives, or teachers being distant? Examine your ideas about stereotyping certain people and professions. While many others may share your impressions, try to find exceptions to the rule. Discuss these stereotypes with your friends to see if your idea of certain people you all know who do not fit in with the prevailing image.

看看你是如何在无形之中可能形成了对他人的刻板印象。你是不是认为所有的运动员都很粗鲁，摇滚音乐人都过着混乱散漫的生活，或者老师都很严肃？想一想你对特定人群或者职业的刻板印象。虽然有很多人可能与你的印象一致，但还是尝试去找到一些例外吧！和你的朋友讨论这些刻板印象，看看有没有你们都认识的人与大众普遍的印象不符。

# 学科知识

## Academic Knowledge

| | |
|---|---|
| 生物学 | 自然界的物种（如鸟类）多样性和巢穴多样性取决于地点和时间的差异；鸟类天生就懂得建造巢穴的技术，而且还在不断完善；雄性动物和雌性之间有着行为、形态和生理方面的差异；雄性动物的颜色是它健康和活力的重要指标；海马的直立姿势是拟态，使它们看起来像海草或珊瑚；海马是肉食动物，能够伏击捕获猎物；海马是有皮肤而没鳞片的鱼，却又不像鱼类那样有尾鳍；海马的眼睛和变色龙的眼睛一样，可各自独立转动。 |
| 化　学 | 侏儒海马的颜色和形状与它所生活的柳珊瑚环境相匹配。 |
| 物　理 | 瓣蹼鹬利用水的表面张力来捕捉食物颗粒，让食物沿着喙到达自己的嘴里；涡旋是一个术语，指的是向下的漩涡；许多鸟类的颜色并不是由色素呈现的；高渗透压环境可以促进海马卵的受精。 |
| 工程学 | 鸟类建筑学描述了鸟类如何设计、施工并且建造巢穴；北京国家体育场俗称“鸟巢”，是世界上最大的钢结构建筑；海马利用软鳍条构成的背鳍向前运动，位于头部侧面的胸鳍负责向左、右、上、下移动。 |
| 经济学 | 干海马作为一种传统中药，零售价每千克600—3000美元。 |
| 伦理学 | 我们需要基于他人对社会的真正价值和贡献来评价他人，而不只是看他们是否符合既定规则；我们对海马的捕捞和鸟巢的采摘（比如燕窝）不能超过它们的再生能力，否则将会导致物种的灭绝。 |
| 历　史 | 大约2 500万年前，海草在西太平洋扩张，之后不久海马就进化出了直立的游泳姿势；海马是希腊、腓尼基人和伊特鲁里亚神话里的形象，它们经常被描绘成上半身是马、下半身是鱼。 |
| 地　理 | 西太平洋的地质构造事件形成了大面积的浅水区域，使海草得以生长，从而导致了海马的进化；海马生活在北纬约52度至南纬约45度的沿岸浅海水域；意大利罗马的特雷维喷泉中有带翅膀的海马雕塑。 |
| 数　学 | 只有不到0.5%的幼年海马可以活到成年；海马卵的重量占雌性体重的三分之一。 |
| 生活方式 | 我们对建设水族馆的渴望会导致珍稀鱼类的过度贸易，从而使它们面临灭绝的风险。 |
| 社会学 | 定式虽然被广泛接受，但所反映的不一定是每个个体的真实特点。 |
| 心理学 | 短视是由于被定势所蒙蔽而无法看到真实的情况，并且没有花时间去仔细观察和分析形成自己的观点。 |
| 系统论 | 珊瑚礁和海草床正在退化，这使海马可用的栖息地大量减少；海马的进出口已经受到《濒危野生动植物种国际贸易公约》的限制。 |

## 情感智慧

## Emotional Intelligence

海　马

海马来到了新的地域，却还在使用旧的思考模式，所以他感到困惑。但他勤学好问，热衷向他人学习更多知识。当海马意识到自己犯的错误时，便很快道歉，然后继续发问，以便更好地了解对他来说新奇的事物——雄性瓣蹼鹬负责筑巢。海马称赞这种鸟是富有爱心的家长。海马对瓣蹼鹬从海底获取食物的技术的简单性感到惊奇。他承认这项技术的有效性，并认为这很聪明。当海马作为一个父亲哺育孩子而得到赞美时，他并没有回应这种称赞，而是悲叹没有更多的父亲能得到这样的赞誉。

瓣蹼鹬

当海马称呼自己为“小鸟夫人”时，瓣蹼鹬并没有生气，只是纠正了海马，没把这事放在心上，他还告诉海马，这个错误是一种定式，当人们没有进行真正的观察时就会发生这种错误。瓣蹼鹬显然很乐意筑巢和照顾雏鸟，并且分享了通过使用水涡旋技术为小鸟寻找食物的简单方法。瓣蹼鹬承认海马在育儿能力方面是独一无二的。当海马抱怨说缺少“特殊爸爸”时，瓣蹼鹬提出了乐观的看法来回应，说世上富有爱心的爸爸比人们想象中多得多。

## 艺术

## The Arts

你家附近的公园或树林里有没有鸟巢？有没有燕子、金丝燕、鹳、织巢鸟或者鸽子？仔细看看不同的鸟巢，观察它们都是用什么材料做的。选择一个主要用树枝和草搭建的巢来研究其建筑构造。你认为你能用手搭建出这样的一个巢吗？现在给你布置一个艺术作业：尝试用草和树枝编织一个巢。记住，你可以用两只手十根手指来编织，而鸟儿们只使用喙！你还可以画出鸟类几种不同的编织技术，尤其是打结技术。

## 思维拓展

### Systems: Making the Connections

我们对世界的认识可能就是一种定式，它让我们形成的观点总是基于标准的信息和知识，而不一定是基于事实。定式能让我们更轻松地在社会中生活，但同时也让我们更难以识别独特的个性、技能和表达。自然界有一个独有的特征，即生物多样性。进化使得地球上产生了各种各样的生命，所以自然界中没有定式。不仅有海马这样一类动物界“真正”的例外，还有着很多各具特色的种类。它们进化出了一些多样化的适应性特征，比如小个头、大肚子、长鼻子、多刺的表皮，以及黑白的色彩。鸟类也在外貌和行为上表现出同样丰富的例外现象，包括有些雄鸟担当起大部分养育后代的家长职责。最有趣的也许是鸟巢多样性及其制造技术，包括打结、编织、隔绝风雨、巢内清洁，当然还有捕猎和获取食物的技术。鸟类和海马的技能以及生活方式给我们提供了新的思维模式。这使得我们始终能在每一个物种——当然还有每一个人——中搜寻并发现独有的特征。当代产业一直在寻求规模经济，这需要标准化，也导致了消费者在品牌和标志的驱动下能轻易识别商品。虽然这样可以降低成本，但同时也导致了对自己独特性认知能力的丧失。这样的社会使人们很少有机会表达自己，以至于许多社会成员默默无闻并感到被社会疏远。如果这个社会能和自然一样允许更多的“规则中的例外”，人们会拥有更大的弹性和灵活性，在压力和困难面前产生更多创造性的应对方法。欣赏多样性不仅仅是欣赏基因多样性和明显的物理特征的多样性，更是欣赏我们的性格和生活方式的多样性。

## 动手能力

### Capacity to Implement

赞美朋友的时间到了，你对一些人的独特看法会使他们感到惊奇。给你的好朋友或尊敬的人列一个表格。找出每个人独特而他们自己又不知道的特点，什么方面都行。也有可能只有你认为这是独特的。用几句话来描写每个人的特点，然后给他们看你对他们的观察和描述。看看他们对于你对他们独特存在方式的赞美是如何反应的。

## 故事灵感来自

This Fable Is Inspired by

# 莎朗·贝尔斯

# Sharon Beals

莎朗·贝尔斯在美国俄勒冈州西雅图的郊区长大，她热爱大自然，但在几十年后才开始拿起望远镜。她自称为“城市自然主义者”。她用她的照片唤起了人们对本土栖息地、河流和海洋的保护意识。

贝尔斯对拍摄鸟巢的形状和其中鸟蛋绚丽的色彩产生了浓厚的兴趣。她的照片展示了鸟儿们巧妙而本能地将树枝、蛛网、虫茧、农作物残余、泥巴、杂物、毛发、苔藓、地衣和羽毛转变成自己养育后代的家园，使我们能一窥大自然非凡建筑师们美丽和神奇的作品。莎朗说：“所有的鸟巢都是鸟儿们用最大的努力用喙和爪子为保护下一代而编制的。”

**图书在版编目(CIP)数据**

冈特生态童书.第四辑：修订版：全36册：汉英对照 / (比)冈特·鲍利著；(哥伦)凯瑟琳娜·巴赫绘；何家振等译.—上海：上海远东出版社，2023
书名原文：Gunter's Fables
ISBN 978-7-5476-1931-5

Ⅰ.①冈… Ⅱ.①冈… ②凯… ③何… Ⅲ.①生态环境-环境保护 儿童读物—汉、英 Ⅳ.①X171.1-49

中国国家版本馆CIP数据核字(2023)第120983号

著作权合同登记号图字09-2023-0612号

**策　　划** 张　蓉
**责任编辑** 张君钦
**封面设计** 魏　来 李　廉

冈特生态童书
**特殊的爸爸**
[比]冈特·鲍利　著
[哥伦]凯瑟琳娜·巴赫　绘
郭光普　译